NON SOLENOID ELECTRIC TRANSFORMER

Akintunde M Lawal

Copyright © 2014 by Akintunde M Lawal
All rights reserved.
ISBN-13: 978-1503347687

PRINTED IN THE UNITED STATES OF AMERICA

This is a non-fiction work. The author has represented and warranted full ownership and/or legal right to publish all the materials in this book.

Non Solenoid Electric Transformers by Akintunde M Lawal

All Rights Reserved.

This book may not be reproduced, transmitted, or stored in whole or in part by any means, including graphic, electronic, or mechanical without the express written consent of the author/publisher except in the case of brief quotations embodied in critical articles and reviews.

Forward

This book is for the reading pleasure of engineers, hobbyists and the general Do-It-Yourself enthusiasts. DIY buffs who wish to insert these new non solenoid electric transformers into appliances should endeavor to consult adept engineers and hobbyists to prevent damage to electric circuits and electrocution when dealing with high voltages. Interested scientists, institutions and individuals wishing to collaborate with the author for further development of this technique should contact the author.

Dedication

To engineers, scientists, hobbyists and do-it-yourself enthusiasts whose quest for simpler and cheaper electric circuit components is insatiable.

CONTENTS Page(s)

Introduction

Principles of Electric Transformers

Principles of NOSET Devices

Uses and application

Future development

Summary

Glossary

References

About the author

Introduction

This book is written for the sole purpose of rekindling a renewed interest in the quest to develop alternative technology to the current crop of weighty electric transformer which make most modern appliance bulky and heavy for consumers.

Electric transformers were developed in the 1880s by American, European and Russian scientists shortly after the discovery of electromagnetic induction. This induction was achieved by making two solenoids around a ferromagnetic solid core.

Types of Transformers

There are different types of transformers depending on the enameled coil windings, primary or secondary solenoid as well as the ferromagnetic core.

Conventional transformers include, autotransformer, audio transformer, output transformer, capacitor voltage transformer, grounding transformer, polyphase transformer, instrument transformer, resonant transformer, leakage transformer, Scott-T transformer and power distribution transformer.

Induction coil transformers do not function with direct electric current (DC) inputs but by alternating current (AC) fluxes only. However, early scientists attempted to use DC power inputs which relied on vibrating electric contacts to run induction coil circuitry.

An attempt is made in this book to introduce a new class of devices aptly named Non Solenoid Electric Transformers which can conveniently substitute conventional transformers in AC or DC electric circuits.

Non Solenoid Electric Transformer (NOSET) neither contains primary or secondary windings nor iron or steel cores. They offer advantages over conventional transformers in their relative ease of construction and readily accessible components.

Unlike the typical transformer, NOSET cannot step up transmission voltages but can conveniently function as cheaper and less noisy substitutes for step-down transformers.

It is envisaged that readers of this book would engage in an in-depth constructive conversation as to whether this new NOSET device is actually a potentiometer rather than a transformer. These devices can be designed as resistors or potentiometers

DIY enthusiasts, adept engineers, hobbyists, industrialists and venture capitalists are encouraged to peruse this book and join the quest to improve its performance and further development.

Principles of Electric Transformers

An electric transformer is a device which transfers electromagnetic flux energy between two or more electric circuits by induction. There is usually a varying or alternating current in the transformer's primary winding solenoid which creates a varying electromagnetic flux in the core and a resonance electromagnetic field in the secondary winding.

Transformers range in size from miniaturized models to colossal entities that transfer electric current in nano-circuits or high voltage transmission lines respectively.

This quintessential device has become the most important appliance for Alternating Current (AC) power generation, transmission, distribution and utilization.

A transformer works only when there an AC power input and its output depends on its primary to secondary solenoid winding ration as well as the value of its input voltage.

A real transformer suffers energy losses which show their limitations from being the ideal transformer.

These losses include, winding joule losses, eddy current losses, hysteresis losses, stray losses, transformer hum, mechanical vibration and noise transmission.

Most transformer cores are made of iron, steel, silicon steel sheets that are bound together as laminations. Powdered iron cores have also being used in some transformers.

Toroidal transformers have also been constructed around a ring-shaped core. Winding are often done using enameled copper wire or Formvar wire.

Transformer expends a lot of heat and high voltage transformers require cooling through the use of mineral oil, paper insulation or convectional air.

Since NOSET devices also be effectively used as resistors, it would be pertinent to mention the functions of conventional resistors and potentiometers that have similar assembly and functions.

Carbon based resistors have been used since the 20th century. These resistors were made of solid cylindrical resistive element (powdered graphite) that was doped with ceramics to modify its conductivity and resistance. The application led to its use in several forms as film, pile and cylinder resistors. However its intolerance to stress, overheating and high voltage and the development of better alternatives has almost terminated the usage of carbon in resistors.

With the advent of nanoscale devices that require very small voltages coupled with the discovery of fullerenes, carbon resistors and devices such as Non Solenoid Electric Transformers and its types of carbon based resistors or potentiometer may make a come back to prominence.

Potentiometer is a multiple terminal resistor that can be used to apply adjustable impedance to the flow of current in an electric circuit. Carbon based potentiometers may be substituted by multiple terminal NOSET devices.

Principles of NOSET Devices

A basic Non Solenoid Electric Transformer (NOSET) is a device which has four embedded wire terminals; two at the edges and two equidistant from the edges of cylindrical solidified granules of a metallic oxide that has been doped with fine particles of carbon to improve its electrical conductivity.

The outer wires at the edges serve as the input terminals while the inner pair wires are the output terminals. Ability of a typical to step down the voltage input can be adjust by varying the metal oxide's carbon content and the distance between input and output wires.

NOSET devices' cylindrical embodiment may contain a single ferromagnetic (such as Iron oxide) or a mixture of many magnetic oxides which are further doped with Carbon granules. Magnetic ceramic or organic semiconductors may also be used as substitutes for the resistive metal oxides. Conductivity of a NOSET device is determined by the quantity of conductive material which is usually powdered carbon used to reduce the resistance of the metal oxide.

So, the more an input voltage is to be leveled to a

lower voltage the less the amount of carbon granules added to the metal oxide.

Rust, Iron II oxide has been used successfully by the author of this book in making the earliest NOSET prototypes. Rust like all other metal oxides is a poor conductor of electricity but it has ferromagnetic properties. An energized atom of rust can use thermal energy to induce electromagnetic induction on proximal atoms.

This principle holds huge importance when rust doped with a resistive conductor like carbon is placed in a field of high potential difference.

Tapping along this field of 'hopping electrons in rust' and easy electron movement among the carbon conductor strands holds the key to the performance of a typical Non Solenoid Electric Transformer device.

NOSET devices step-down AC voltage input to AC voltage output or DC input to DC output.

The electrical conductivity and step-down capability of any NOSET device can be adjusted by doping the metallic oxides with fine granules of graphite, buckminsterfullerene or other organic conductors. It should be noted that NOSET devices may lose its stability and performance due to their being subjected to high voltage or overheating.

Uses and application

NOSET devices are cheaper and easily constructed substitutes for step-down transformers in low voltage circuitry and appliances.

They function with AC or DC power inputs. NOSET components are readily available, accessible, easy to construct or manufacture.

They can be designed as two terminal resistors.

They can also be designed as multiple terminal potentiometers.

The electromagnetic fluxes they generate and performance parameters can be easily monitored and deduced.

They assist in minimizing power lower by their resistivity.

They offer cheaper and easier to construct alternatives to carbon resistors and potentiometers.

These devices can be easily miniaturized or enlarged to perform various tasks from nanoscale devices to colossal electrical step down transformers if a more doping stable substance like carbon is developed. Would carbon fullerenes fulfill this role? That's the main puzzling question this book wants to draw the attention of high profile research scientists and industrialist to.

Future development

Compact manufacturing of NOSET devices for easy handling, insertion into electric circuitry and usage.

NOSET could also be developed into various types of transformer classes such as:

Audio transformers which obstruct radio frequency interference.

Autotransformer for stepping down high voltages.

Instrument transformers to step-down voltage to a low value which is standardized for an instrument.

Polyphase transformers which has more than one output that power multiple circuits.

Color-coding of NOSET devices for standardized recognition of the performance of each device.

Scientific data and specifications for comparative performance of NOSET devices with conventional electric transformers.

Analysis of performance of miniaturized or large NOSET devices.

There is a need to prevent NOSET instability due to very high voltage and over heating.

Search for a thermo-stable alternative to carbon as the NOSET conductivity improving agent in order to improve their performance and stability over time.

Development of printable NOSET devices on hybrid PCB modules.

Development of thick or thin film NOSET devices for miniature electric circuits.

Limitations

The use of Non Solenoid Electric Transformers may be restricted to the ability to reduce voltage inputs only. Hence, it cannot be used when increase of input voltage is desired.

The devices may not be able to withstand high operational temperatures like conventional solenoid transformers. Therefore, their use will be limited to low voltage appliances unless scientists can develop a substances that has the conductive, piezoelectric as well as resistivity properties of carbon but is more stable when subjected to stress, overheating and very high voltage.

Summary

Non Solenoid Electric Transformers are devices made principally of metal oxide and trace carbon granules. The doped semiconductor metal oxide has four terminals of embedded wires with each of the two pairs of terminals acting as input and output outlets respectively.

These devices can perform the task of stepping down high voltage inputs in electric gadgets or appliances that required lower voltage for their optimal performance. Scientists should endeavor to join the quest to improve the performance and stability of NOSET devices because they can function in DC or AC circuitry. They are easily constructed or manufactured from relatively cheap and readily available components. These devices can also be used to develop resistors and potentiometer.

Since carbon fullerenes have been found to have heat resistance and super conductivity would they make more stable NOSET devices a reality?

Glossary

Transformer is a device in which varying input electric current induces a lower or higher output electric current depending on the windings ratio of the solenoids involved.

Resistor is an electric device which reduces the flow of electric current.

Solenoid is a coil which is wound into a closely packed helix for the purpose of generating electromagnetic flux.

Induction coil comprises of two or more solenoids that are wound around or proximal to a common core.

Enameled wire is a conducting wire that is covered by a very thin layer of insulator polymer substance.

Ferromagnetic substance is a material which can produce its own magnetic field or be induced to produce a magnetic field.

Core is a solid ferromagnetic substance which produces a uniform magnetic field within a specific volume of space around the solenoid.

Potentiometer is a multi-terminal resistor with a sliding or tunable and adjustable voltage divider.

References
1. Crosby, D (1958) "The Ideal Transformer" IRE Transactions on Circuit. Theory 5 (2): 145

2. Heathcote, Martin (Nov 3, 1998). J & P Transformer Book (12th ed.). Newnes. pp. 2-3

3. Winders, John J., Jr. (2002) Power Transformer Principles and Applications. CRC. pp. 20-21

4. Kulkarnic, S.V.; Khaparde, S. A., (2004) Transformer Engineering: Design and Practice. CRC Press.

5. Cornell, RM; Schwertmann, U (2003) The iron oxides: structure, properties, reactions, occurrence and uses. Wiley VCH.

6. Deprez, N.; McLachan, D. S. (1988). "The analysis of the electrical conductivity of

graphites, conductivity of graphites powders during compaction." Journal of Physics D; Applied Physics (Institute of Physics) 21 (1): 101.

7. Tipler, Paul (1998) . Physics for Scientists and Engineers: Vo.2: Light, Electricity and Magnetism (4th ed.). W.H. Freeman.

8. Sommer, T; Kruse, T; Roth, P (1996) "Thermal stability of fullerenes: a shock tube study on the pyrolysis of C60 and C70." J. Phys. B: At. Mol. Opt. Phys. 29 4055.

About the author

Akintunde M. Lawal is a freelance scientist, with a postgraduate degree in the sciences. He is the author of a Sci-Fi novel titled 27th Century Fiasco and inventor of four mechanical devices.

Contact: akintundemlawal@gmail.com

Watch out for series of demonstration projects caption 'Non Solenoid Electric Transformers' on YouTube very soon.

www.ingramcontent.com/pod-product-compliance
Lightning Source LLC
Chambersburg PA
CBHW070734180526
45167CB00004B/1746